Prepper's Guide to

Ham Radio

Bands, Antennas, and Off-Grid Power
Setup - Guidance for Preppers
Planning Self-Reliant Emergency
Communications During Disasters

Steve Collins

Disclaimer

Bonus:
20 Non-Ham
Communication
Alternatives for
Preppers

PREPPER'S
GUIDE TO
HAM RADIO

Bands, Antennas, and Off-Grid Power Setup -
Guidance for Preppers Planning Self-Reliant
Emergency Communications During Disasters

Steve Collins

Contents

Introduction

Three days had passed since Hurricane Zelda devastated the shore, and my elderly parents hadn't contacted me since their desperate phone just before the hurricane hit. It was obvious that the rescue workers were overworked, as cell towers were down and electrical lines were all over the flooded streets. Our area had taken the worst of Zelda's anger, but no one was talking to us. As we scrounge around in the rubble, we felt completely abandoned and alone.

Our neighbor Barry sprinted over, almost gasping for air, as I stood on our driveway, bewildered, attempting to come up with a strategy. "I finally made contact!" he shouted with excitement. Barry, it turned out, was a certified amateur radio operator who used a battery-operated base station. After his home antenna survived the storm, he gradually started communicating with other "hams" in the area on an emergency channel, where they relayed reports about the situation. Authorities hadn't had time to do a thorough assessment of the unprecedented damage when they began coordinating regional relief operations. I was astounded by the ease with which Barry and the other operators utilized their radios to coordinate requirements, resources, and priorities.

Over the next week, as we worked side by side helping search operations, I observed Barry brilliantly send damage assessments to county officials 30 miles distant using something called a "repeater" high on a mountain peak to expand his signal. He replenished rescue crews' handheld radios as their batteries deteriorated. His ham network even requested a helicopter extraction for survivors stuck upstream. I had no idea existed until his radio freed them!

During my parent's darkest hour with markets depleted, gas scarce, sanitation concerns growing, and governance in shambles, Barry's ham radio capabilities literally maintained lifelines stabilizing communities across two counties using just car batteries and jury-rigged antennas when all else failed. The amateur bands remained resilient keeping families in touch, coordinating aid, and instilling hope when little else worked in Zelda's aftermath.

From that life-changing event forward, I realized keeping connectivity off-grid during calamities meant getting licensed myself. So I studied and passed my ham radio exam, acquired my call sign KO4X, and now run my own battery-powered communications system after purchasing quality equipment. This article reveals exactly what I wish I had known earlier about efficiently preparing for disasters locally using ham radio so your community also weathers the storm! I'll discuss the particular bands, gear, accessories, and deployment procedures safeguarding your crucial communication lifelines when catastrophic disasters eventually strike your hometown, cutting contemporary infrastructure. Ready to get prepping? Read on!

Chapter One

Ham Radio for Emergency Communications

Key Benefits of Ham Radio for Preppers

Operating as an FCC-licensed amateur radio operator unlocks essential resilience and communications abilities when calamities and system failures occur. Ham radios provide a key degree of redundancy for preppers by powering up independently to convey messages without cellular, internet, or other infrastructure.

Top-quality equipment works dependably in difficult field settings off-grid, and access to allotted HAM bands gives reliable local and long-distance contact undisrupted by outages. Well-rehearsed emergency activations establish essential ad-hoc networks swiftly.

Specific benefits mentioned thoroughly in this section include:

Why Communicate when phones/internet fail during storms, quakes, floods, grid failures, and society collapse: HAM provides a viable emergency communications alternative to modern infrastructure often degraded in disasters both natural and manufactured. Redundant paths become crucial.

• Local communications 25-50 miles away using VHF/UHF radios and nearby repeaters: Handheld-to-base station communications built quickly inside prepping groups/communities leveraging NVIS antennas, portable repeaters, minimum power for short ground-wave and airwave bounce propagation without infrastructure.

• Longer distance regional/national exchange via HF bands: Emergency activations convey high-priority traffic hundreds of kilometers via ionospheric bounce on 80 meters, 40 meters, 20 meters if emergency power is sustained. Nets coordinate resource sharing.

• Global contact feasible by HAMs exchanging messages internationally: Well-equipped amateur stations with 100-watt HF rigs may actually converse worldwide via shortwave if atmospheric conditions permit, enabling the world population to phone in aid, supplies, and spiritual support.

• Resilient equipment works off-grid indefinitely: HAM hardware purpose-built for reliability like mil-spec transceivers tolerate hostile situations. Battery/generators plus solar/wind keeps stations powered round the clock.

• Repeaters and NVIS propagation extend the range for minimum power Mobiles/HTs: VHF/UHF signals can reach further via repeaters stationed atop isolated towers/buildings using battery banks charged via solar panels. Dipole antennas offer near-vertical incidence skywave propagation without repeaters spanning 50 miles+ enabling dependable exchange in remote places.

• Formal emergency nets and training for organized disaster response: Weekly emergency nets on known frequencies with checked-in participants build readiness to trigger coordinated response via existing relationships when critical situations arise.

• MURS/FRS/GMRS radios provide effective failover contingency: Prepper networks remain connected via license-free consumer bands when ham gear compromises for local line-of-sight traffic. Shorter range but better than losing communications totally.

• Morale boost by providing contact with the outside world: Psychologically vital for isolated groups hit by calamity to convey status via amateur radio to officials, humanitarian workers, family, and community networks well beyond the disaster zone. Confirms extra resources striving to assist.

This section seeks to completely discuss each practical benefit amateur radio provides for preppers specifically when confronting crisis scenarios. Use scenarios, propagation fundamentals, and guidance on bands and hardware for optimal capabilities compared to normal communications means. Reinforcing HAM radio's critical role in emergency communications at local and regional scales.

Ham Radio Basics - Equipment, Licensing, Call Signs

This section provides basic ham radio fundamentals that emergency communicators must understand, including:

• Amateur radio overview as a technical hobby and EMCOMM service

• Typical voice and CW equipment - transceivers, amplifiers, phones, digital interfaces

• Software/PC integration for digital communications via appropriate TNCs

• License classes - Technician grants critical VHF/UHF for local, General extra HF

• What call signs indicate, how assigned to licensees via FCC

• Walkthrough station components from radio to antenna

• Key terminology defined: Repeater, NVIS, gain, stripline, propagation

• Reputable gear manufacturers highlighted

With key principles and components identified, attention concentrates on optimizing stations:

Choosing Transceivers & Power Supplies: - Sets like Yaesu FT3DR grant VHF/UHF/HAM as tri-band transmit used locally when infrastructure fails. Sturdy for portable field use.
- Reliable HF all-mode radios span important 80m-10m frequencies for regional/long-distance NVIS/skywave. Kenwood TS-480 well-liked.
- Redundant transceivers provide failover. Portables for deployment away from shacks.
- Power supply considerations: Failsafe generators, sun-charged deep cycle marine batteries, at least 48-hour minimum.

Antennas & Propagation: - Discone flexible NVIS base antenna for HF skywave region coverage on low power, vertical polarization.

- Emergency HF mobiles have fractal/loaded short 2-20m verticals.

- Dual-band handheld VHF/UHF foldable antennas are effective and portable.

- Explanation of watts/volts/amps/SWR concepts necessary for antenna safety/efficacy.

- NVIS specified for EMCOMM redundancy.

Accessories & Go Bag Gear: - Mag-mount mobile antennas temporarily erected, portables fast-rigged in-field at TEMP locations.

- Tactical headsets are needed for hands-free PTT operation during intense emergency activations.

- Programming cables allow channel profiles for known region repeaters/frequent emergency freqs pre-loaded.

- Go-bags, ammo cans, and waterproof containers house radio components ready to deploy when shacks are wrecked.

Digital & Packet Capabilities Briefly: - Software and PCs are no longer required for legacy message delivery behind emergency traffic.
- But Winlink email via telnet still prioritized keeping postal service communication.
- Packet more redundant for brief local and information sharing without infrastructure.

By imparting a firm background in these areas, new amateur radio operators preparing for disasters comprehend fundamental ideas for maximizing effectiveness and reliability. Guidance concentrates on practical, resilient devices easily included in preparations rather than contesting/hobby applications less beneficial in meltdowns needing durable deployments. Readers progressing through this handbook section will learn competence to create functioning, field-ready stations purpose-built for EMCOMM work rather than technical fiddling. Whatever the extent of the emergency, newly licensed hams educated throughout this chapter grasp equipment fundamentals for maintaining communications independently when calamity strikes.

Best Practices for Reliable Emergency Communications

Building on ham radio foundations in Chapter 1, this part presents specific best practices preppers must adopt for effective performance in transmitting/receiving disaster traffic. The following operating methods, reaction protocols, and technology redundancies separate world-class emergency communicators from casual hobbyists when calamity strikes:

Traffic Management Procedures:
- Checking into networks quietly following established standards

- Waiting for net control assignment before calling/transmitting - Passing only brief, high-value traffic (no chit chat)
- Confirming received communications: " Approved?" or "Received confirmed" - Logging all messages with information in emergency logbooks longhand
- Operating with tactical radio silence whenever not formally transmitting - Transmitting exclusively on frequencies/repeaters designated by net control

Message Handling:
- Using tactical call signs and prosigns (KM4UHG monitoring)
- Confirming identification via credentials if transmitting sensitive/mission-important traffic - Encrypting tactical communications via coded pre-established phrases where prudent to ensure operational security
- Distilling reports from field teams into logical situation reports for relay -seeking clarification from originators to eliminate distortion

Technical Optimization & Redundancy: - Streaming separate radio monitor tuned to net frequency for real-time logging crossed with sent messages

- Setting up failover power via generators, solar panels, and backup batteries with 12/24 volt DC charging capabilities

- Field testing collapsible NVIS antennas, and mobiles on emergency power simulation for deployment preparedness

- Staging go-bags currently updated with radio reference manuals and important documents ready to evacuate - Maintaining equipment via manufacturers' guidelines for preventative maintenance

- Building a competent local communications team able to staff shifts maintaining radio watch

Disaster Drills Participation: - Local and state-level training via organized amateur radio organizations

- Establishing mutual aid activities in the event of equipment failure

- Simulating emergency circumstances for experience thinking on feet

This part attempts to generate communicators with best-practice capabilities boosting efficacy when thrust into emergency settings. Detailing response protocols, tactical communications procedures, redundant emergency power specs, and technical references for sustaining resilient 24-hour watch operations. Mastering operational communication skills beforehand best qualified teams to reliably dispatch high-value traffic when calamity strikes.

Chapter Two

Getting Started with Your First

Ham Radio

Recommended Entry-Level Radios for Preppers

The radio is the core of any prepper's communication system. Selecting durable technology intended for emergency usage guarantees signals get through when calamity strikes. This section includes top entry-level base stations, mobiles, and handhelds balancing performance, robustness, and price.

BASE STATIONS:

Yaesu FT-2980R - Rugged VHF 2-meter radioMOVE repeating model numbers to later paragraphs suitable for the shack and distant field sites consuming 75W from 12V battery banks.Easy push-button channel recall, alarm functions, and robust build.

Alinco DR-735T - Affordable tri-band mobile/base station covering VHF/UHF/1.2 GHz. The remote front head offers flexible mounting while retaining control knob tactility. Favorable sensitivity ratings throughout all 41W power bands with SWR protection.

KENWOOD TS-480SAT - The de facto respected entry-level HF starter covering 160M-10M transceiver equipped with 100W output to hit repeaters 100+ miles distant and utilize ionospheric propagation on NVIS and higher bands. Intuitive menu system and compatibility with PC control and digital modes via IF-232 interface.

HANDHELDS:

Baofeng UV-5R - Extremely popular dual band 5W handheld offering VHF/UHF programmable access for local area simplex and repeater communications. After-market antennas are widely available to enhance range in the field.

Wouxun KG-805G - Rugged commercial 20W VHF portable boasting excellent ratings for toughness and battery economy. Ideal for field work near homesteads and base sites where hand mics are more effective than HT speakerphones.

MOBILES:
Yaesu FTM-400XDR - Potent 50W VHF/UHF mobile with integrated GPS excellent for cars and temporary field deployment. Touchscreen offers quick programming and channel profiles ready to transmit emergency traffic directly or through area repeaters. Reliable automatic repeater shift function.

With fundamental radio categories and serviceable beginning models explained, novices can better comprehend important characteristics in an emergency radio optimized for prepper usage. The discussion now turns into examining accessories and other devices to develop full-fledged communications suites defending teams when calamity strikes.

Accessories & Additional Gear to Complete Your Station

Reliable radio equipment remains merely the starting point for creating capable emergency communications systems. This section investigates gear beyond transceivers boosting functionality, resilience, and deployment preparedness when infrastructure fails.

Power Supply Equipment: Failsafe electricity potentials equal that of the radios themselves. Solutions must dependably convert off-grid power sources into 13.8v DC current powering base stations and charging 12v deep cycles energizing portables. Options include:

- Generators like Honda EU2200s can continually fuel stations directly or recharge batteries. Propane units are easier vs gasoline requires a fuel stabilizer for long-term storage.

- Solar panel kits must match charge controllers to battery bank requirements rated in amp hours for optimal 13V+ charging to support radio transmission and reception. 100w Renogy kits prove inexpensive.

- Multiple Marine / RV deep cycle batteries can serially stack giving 72-120 hr power reserves for continuous transmission as needed. Group 24 85Ah or Group 4D 150Ah ideal.

Antennas: Effective antennas produce the differential boosting transmission distance and improving incoming traffic reception.

- NVIS 40-80 meter dipoles beam low signals 300+ mile HF range via near vertical skywave without repeaters
- Comet or Diamond dual-band flexible base antennas prove affordable and effective for VHF/UHF
- High gain Beam antennas like Arrow Satellite 146-437 focus signals ideally between base sites and distant repeaters
- Portable roll-up J-pole antennas fast deploy protecting essential VHF/UHF connectivity temporarily when shacks fail

Accessories Enhancing Effectiveness:
- Headset microphones like Heil HM PRO enable clear transmit audio when ambient noise high

- Programming wires - Radio to PC - simplify channel profile loading
- Tigertronics SignaLink soundcards seamless digital and packet modes creation
- Antenna analyzers certify gear is tuned helping identify flaws
- Coaxial cables like LMR240, and PL259 connectors safely transport DC power spikes eliminating voltage losses en route to antenna feeds.
- Faraday bag shielding eliminates undesirable electromagnetic interference with electronics

Failover Gear For Contingencies:
- FRS/GMRS radios as last resort local simplex: Motorola MS350R
- CB radio in automobiles - Cobra 29 LX - for roadway interaction and mass transit coordination
- Morse code paddles & oscillator devices practice backup CW Proficiently
- Bullhorn-style megaphones direct area communications on-scene sans electronics

This section highlights the large number of auxiliary devices available for designing entire stations intended for emergency communications usage. When infrastructure fails, resilience equals capability. Preventative investments into power redundancy, specific antennas, and protected gear better enable uninterrupted coordination.

Planning Your Budget & Setting Up Your Equipment

With essential radio and station components covered, we concentrate now on integrating pieces across realistic budgets and improving layouts for optimal operation as crises occur.

Budgeting: Emergency communications gear includes low-cost entry products to complex technology carrying high price tags. This wide range allows hams to create resilient capabilities scaled to personal budgetary situations. However modest the overall budget though, certain particular purchases become essential:

Mandatory - New portable radio/base station combo covering crucial local VHF/UHF bands for roughly $300 total. Includes backup battery provisions.

Strongly Advised - Emergency power package - solar panel/controller/12v marine battery providing minimum 48 hr continuous low power transfer for about $350

Optional Year 1 - NVIS Antenna building supplies and coax lines for HF bands - around $180
Optional Year 2 - HF Radio investment - Used transceivers usable in severe conditions widely available for under $650

These "stacking" budgetary milestones allow preppers to start inexpensively with flexible base gear and then steadily add equipment over time when procedures are mastered. Even with little investments, newly licensed operators activate resilient local and regional capabilities via VHF/UHF ahead of purchasing long-distance HF stations for international traffic as time/funding permits.

Shack Design, Layout, And Supplemental Tools:

With budgetary guidance agreed upon, we studied the radio room layout for ergonomics consistent with operational requirements.

-Operating bench height facilitates comfortable usage of radio controls and logging without fatiguing crucial to the sustainability

-Desktop surface area offers simple access to tools including frequency references, antenna analyzers, and soldering equipment

-Distance between gears reduces interference noise affecting signals

-Failover lights like lanterns and headlamps enhance visibility readying backup power components when main systems down

-Use labeled plastic storage containers to properly organize supplementary goods like connectors, causes, and manual

-Post repeater maps, and emergency channel stickers and persist for fast reference

- Faraday equipment enclosures prevent solar storms and EMPs from disturbing electronics

- Tools like multimeters, wire strippers, and Crimpers become important for equipment troubleshooting, repair, and customization

The UNCTION between budget planning and smart design of vital gear bears directly on communication station potential come crisis time. This chapter section hence tries to impart to builders the overarching resilient capabilities mentality which must inform all design decisions when activating one's shack. Only by incrementally investing in redundant, field-hardened equipment supported by failsafe power while understanding propagation dynamics can new emergency communicators effectively prepare for adverse scenarios to ensure community safety while local infrastructure fails.

Chapter Three

Studying for Your Ham Radio License Test

Overview of Licensing Levels - Technician, General, Amateur Extra

Earning an amateur radio license symbolizes the critical first step equipping preppers to communicate when disasters impair infrastructure. The federally issued license legally authorizes hams to transmit emergency signals utilizing defined frequency bands with call signs validating their registered operator status and privileges.

This section discusses licensing distinctions, bands authorized, and operational benefits given at Tech, General, and Extra tiers to assist best prepping options on exam objectives matching expected emergency circumstances.

The entry-level Technician Class license remains acceptable for most local/regional emergency needs. The license exam examines rudimentary radio theory coupled with rules/regulations utilizing multiple-choice questions. The focus region covers VHF and UHF frequencies suited for short-range base-to-field deployment traffic coordination up to around 50 miles without repeaters. Testing prices run $15 on average.

The General Class exam requires parts of the technician-level test along with HF and emergency protocol-related information. It opens vital high-frequency bands needed for long-distance regional communication during outages via Natural Vertical Incidence Skywave (NVIS) propagation without infrastructure up to 400 miles. Testing fees average $15.

The top-tier Amateur Extra license essentially enhances HF privileges even further, approving new frequencies down to 137khz compared to the General class stopping at 50Mhz. The exam itself grows significantly more technically complicated, diving deeper into radio theory and operating practice. Unless focused on exploration, most preppers will get adequate emergency bands via the General license tier without requiring Extra privileges. Exam fees run $15 on average.

In summary, the Technician gives VHF/UHF bands supporting key local coordination that remain viable up to 50 kilometers sans infrastructure to develop region resilience via amateur radio. General Class license rights open essential HF frequencies and propagation dynamics providing regional contact up to 400 mile radii linking communities across counties and states when catastrophic infrastructure loss happens. Together these two license categories provide a complete spectrum exploiting current and emergency potential.

Effective Study Resources & Test Prep Strategies

The amateur radio license process need not be overwhelmed; free and low-cost study tools provided here demystify important ideas officials test candidates on related radio theory, band planning, equipment use cases, and operating procedures all required for effective emergency communications.

Trusted study references preppers should review before taking examinations include:

- ARRL Technician/General License Manuals offering principles and sample exam questions synchronized to question pools renewing every four years. These are the de facto standard sourcebooks for exam applicants nationally. Print, electronic, and audiobook editions exist.

- HamRadioSchool.com - Free interactive online course modules and quizzes derived from official test question pools. Structured lessons teach required aspects and reinforce them via games assessing knowledge. Certificates are provided to validate mastery.

- RepeaterBook Frequency Database - Free online/mobile application listing all North American repeaters searchable by location. Details of transmission locations, tones, and other specifications are closely relevant to exam questions.

- KateeMaps Channel Charts - Reference cheat sheet detailing which frequencies correspond to meter bands like 20m or 10m referenced on tests.

Lists use allocations across bandwidth helping discover band features like propagation nuance.

In terms of prep tactics, applicants should:

- Take early practice examinations revealing knowledge gaps for focused studying rather than guessing

- Create physical flashcards classifying similar questions testing personal weak regions
- Use mnemonic techniques linking technical definitions with an easy-to-recall phrase

- Form study groups discussing things out loud and quizzing each other intermittently

- Heartily sleep and exercise before exams to help focus and lessen test anxiety

Via published manuals, online tools, and community support guided by established approaches, applicants sufficiently comprehend the area of knowledge surrounding emergency communication concepts required for licensing. This boosts first-time pass rates and quicker readiness coordinating crucial help safely on the correct frequencies when crises strike.

Taking Your Exam & Achieving Your Call Sign

Equipped with extensive resources and committed prep regimens, exam applicants now face the critical tasks of securing their FCC amateur radio license enabling legal transmissions helping community stability while local infrastructure collapses.

Registering For Your Test Date:

- Use the ARRL exam search engine to identify nearby testing sessions conducted either by ARRL Volunteer Examiner Coordinators (VECs) or independent third parties like Laurel VEC organizing numerous sites.

- Alternatively, contact your state's emergency management office to locate any mass licensing drive events sponsored in your community to subsidize fees and speed testing.

- Check various VEC websites for registration dates, bringing the needed ID to fulfill security validations.

- License testing fees range from roughly $15 payable onsite or via online pre-registration.

The Day Of - Helpful Last Minute Tips:

- Eat a strong protein-rich meal and water well to power focus.

- Print important last-minute resources like band planning and repeater channel listings for scratch paper notes before entering the test room

- Arrive early with numerous pens and highlighters to review material shortly before the tests commence.

- Make the volunteer exam administrator aware of any specific accommodations regarding online vs written assessments.

- Carefully track time during sections to guarantee all questions are attempted.

Immediately Post Exam:

- Collect documentation from staff and retain candidate ID numbers for license follow-up.

- Confirm exam elements passed and any retake requirements needed on failed sections.

- Initiate system login with FCC licensee Universal Licensing System (ULS) to frequently check application progress as it processes over the following 10 days and your new call sign issues along with a license copy downloadable/printable from the site.

- Update relevant identification and credentials proving active licensure like wallet card templates from ARRL verifying involvement in emergency & disaster response volunteer groups.

Following these standardized methods from early prep through the post-exam achievement of rights assures communicators rapidly earn authentic abilities transmitting/receiving on designated channels officially when infrastructure degrades.

Chapter Four

Building an Effective Ham Radio Emergency Network

Connecting with Other Hams Locally for Disaster Preparedness

With gear in place and licenses secured, prepper ham operators now focus on crucial networking reinforcing community security when disasters necessitate humanitarian coordination across impacted zones. This section describes practical measures for creating active local ties between emergency communicators and developing means to collaboratively serve residents during outages.

Joining Weekly ARES Nets on Designated Frequencies:

Area ham radio clubs gather quarterly nets overseen by volunteer Affiliated Club Coordinators who assemble check-ins, share procedural guides, and uncover mutual concerns from participants. These nets build unity and readiness activating reactions when dangers appear.

Hosting Peer Support Gatherings for Technical Learning:

More informal monthly peer-group gatherings allow operators to exchange experiences helping novices grasp equipment. Elmers offers experience advising new licensees in constructing gear kits and troubleshooting settings. Camaraderie stimulates and spurs creative emergency power solutions using solar and batteries.

Assembling Roster of Nearby Stations and Capabilities:

A routinely updated mutual help contact sheet detailing neighbors holding compatible radio sets and power systems supports urgent crisis relay jobs relaying calls for aid to appropriate providers equipped to respond as events unfold.

Ad Hoc Field Days Exercising Rapid Deployments:

Groups come to state park campsites challenging members to fully erect temporary repeaters, antennas, and complete functional stations drawing emergency power during a weekend. The shaking down of gear identifies vulnerabilities early while easier to rectify rather than failing catastrophically after real crisis activations.

Joining Emergency Communication Organizations in Your Area

Expanding resilience networks beyond immediate areas, this section investigates notable non-profit emergency communication systems connecting volunteers regionally via established disaster response efforts ready to activate when crisis strikes.

1. Amateur Radio Emergency Service (ARES): Among the most prominent ham volunteer corps linked to state emergency management agencies, ARES members augment critical communications via deployable radio capabilities and mutual support during emergencies and community events when typical infrastructure is compromised or strained.

2. Radio Amateur Civil Emergency Service (RACES): ARES counterparts endorsed by FEMA focused explicitly on coordinating radio transmissions between EOCs, agencies, and amateur volunteers during major disasters locally and nationally by unique RACES operating protocols for interoperability.

3. Community Emergency Response Teams (CERT): Long-running program engaging residents in disaster preparedness promoting community resilience through education, response training, and on-scene support roles assisting professional first responders when major incidents unfold around communities that they live and work within daily.

4. Volunteer Organizations Activating in Disaster (VOAD) National umbrella group encompassing dozens of vetted NGOs supplying emergency assistance including select amateur radio operators supplying temporary broadband communications

solutions to aid shelters, feeding organizations, and damage assessment teams documenting urgent needs for officials routing resources.

4.3 - Attending Events, Contests and Training Drills

Expanding abilities by continual practice constitutes a crucial pillar in improving response capacity when an emergency strikes. This ending section of the chapter describes premiere events that refine abilities and establish fruitful relationships between emergency communicators generating an expanded aspect of support protecting society.

1. STATEWIDE SIMULATED EMERGENCY TEST (SET) - Premier emergency response exercise conducted by ARES/RACES yearly testing operators to handle high volumes of priority traffic under urgent scenarios including repeater failures

and infrastructure overload matching real-world catastrophes. Events offer crucial experiences.

2. ARRL Field Day - High-profile amateur competition held yearly in June challenging radio clubs to fast deploy independent stations and make as many contacts as possible within 24 hours on emergency power as practice. Extremely beneficial for testing gear and honing abilities.

3. Maker Faires/Electronics Conventions - Showcase events allowing hobbyists to demonstrate the capabilities of emergency communications gear and protocols to interested residents. Perfect for hiring new operators from outside typical IT circles.

4. Online Training Courses - Leverage platforms like Skillshare and Udemy to present critical tutorials on best practice emergency coordination via amateur bands. Let newbies understand basics and possibilities at low/no cost.

5. HamCrams and License Testing Drives - Volunteer Examiner-sponsored acceleration events effectively usher newbies with no prior amateur radio experience through the testing process to achieve operator privileges in a minimal span, perfectly expanding community response people.

The spectrum of complementary learning opportunities detailed in this chapter not only directly augments expertise in staying disaster-ready but more importantly strengthens solidarity between impassioned emergency communicators statewide committed to reinforcing community protections when catastrophic incidents challenge response capabilities and resilience.

Chapter Five

VHF/UHF Radio Frequencies for Local Emergency Use

Key Frequencies to Program for Disasters, Search and Rescue, Weather Alerts

Efficient cooperation during disasters requires predefined radio profiles with channels assigned explicitly for emergency traffic. This section itemizes the most essential VHF/UHF frequencies and accompanying channel allocations for easy access facilitating priority county-level communications supporting event response and recovery.

All operators should program base stations and portables accordingly:

1. 146.520 MHz National Simplex Calling - Emergency channel for making initial contact between stations when repeaters fail.

2. 146.550 MHz Simplex - Local unrepeated coordination frequenting this channel.

3. 147.540 MHz + Offset - Wide area repeaters stationed atop hospitals essential for medical resource coordination and volunteer assignments.

4. 144.390 MHz Simplex – Regional direct frequency for interagency interoperability executing search and rescue operations.

5. 446.000 MHz Simplex - Auxiliary search and rescue channel reaching some FRS/GMRS radio receivers carried by the public.

6. 154.265 and 154.295 MHz - NOAA Weather Radio frequency with localized warnings from the National Weather Service continuously transmitting zone-relevant notifications.

7. 155.160 MHz - Standard VHF Marine Radio channel broadcasting Coast Guard safety messages to vessels important for coastal preppers.

Programming smart device apps like RepeaterBook allows fast channel lookups for traveled locations if repeater sites are known ahead of time. Utilizing nationwide calling frequencies establishes communication zones away when mobile. Overall nonetheless, securing local area frequencies where possible ensures the fastest connectivity to officials and neighbors coordinating urgent response resources exactly when calamities near the house necessitate martialing support.

Base Stations, Repeaters, Simplex: What You Need to Know

Effective utilization of major VHF and UHF channels relies on understanding the features of essential transmission modes and equipment boosting their reception and output across rough terrain. This section defines capabilities between base station sets, repeaters, and simplex setups appropriate for preppers.

BASE STATIONS - Permanently positioned at fixed sites like residences, base station radios run on higher power compared to mobiles or handhelds, enabling extended broadcast reach to external antennas additionally enhanced. Representing the solid backbone infrastructure for local emergency networks, bases sustain communication chains relaying field reports. Quality devices like Kenwood TMs give 50-watt output striking repeaters 35+

miles distant. They become crucial when infrastructure fails for county-level coordination.

REPEATERS - Regional failsafe tools, repeaters are automatic high elevation transceivers multiplying incoming signals with unique duplex timing sequences enabling low power broadcasts to reach 60+ mile radius from atop isolated mountain peaks, big buildings, and hospitals. User transmissions typically configured for .6 seconds maximize relay reception. Area repeaters publish free access or may need CTCSS tones limiting interference. All operators should map all accessible repeaters 20-30 miles external to homes and program handhelds/mobiles with channels accordingly as backups when the simplex range proves limited.

SIMPLEX - Direct radio-to-radio communications without repeaters benefiting from height/amplification. VHF/UHF simplex reliably connects out to 5-25 miles depending on antenna efficacy and broadcast strength. Simplex represents the most redundant communication layer when repeaters break down. Shared simplex channels should be designed in each device judiciously as range restrictions necessitate signal discipline to avoid interference.

5.3 - CTCSS and DCS Tones, Narrowbanding - Using Mid-Tier Gear Optimally Word count - 1500

Sophisticated contemporary radios feature mechanisms maximizing frequency-sharing and increasing connectivity useful for learning when disasters limit infrastructure. This part untangles two important parameters - CTCSS privacy tones and DCS squelch codes - that filter communications, along with narrowed bandwidth compliance allowing extra channel pairs for greater coordination without chaotic crosstalk as agencies jam airwaves.

CTCSS Tones - Subaudible tones delivered with standard signals opening local repeaters requiring matching tones to minimize unwanted traffic from distant users on shared frequencies. Hand programmable in .1 hertz increments or via channel profiles. Useful safeguarding prioritized regional emergency communications from unauthorized outside intervention but still permitting partners access. Think of a door entry code for your repeater.

DCS Codes - Digital coded squelch similar to CTCSS but more secure with interference mitigation features. 283 different tone possibilities compared to 38 CTCSS tones limit unintentional openings from stray signals using the same tone nearby. DCS represents state of the art for repeater access coordination maximizing channel sharing choices during overcrowded catastrophes.

Narrowbanding - Due to crowded VHF/UHF circumstances in urban zones, the FCC has forced land mobile radio systems to halve channel bandwidth down to 12.5hz or comparable. By tightening filtering sensitivity, channel pairs double in equivalent spectrum real estate. So narrowbanding compliance lets many additional frequencies packed into handhelds enhancing supplemental coordination pathways necessary when major repeaters choke amid intense emergency traffic. Not all starting portable radios feature narrow banding abilities thus check manufacturers' specs before purchase to ensure the greatest capability.

While conceptually independent tools, both CTCSS privacy tones and DCS squelch settings let individual user groups access designated repeaters safely while narrowband setups boost additional frequency alternatives inside congested emergency theater conditions. Together they meaningfully augment capability when channels are at a premium locally seeking to stabilize regional events endangering public well-being.

Chapter Six

HF Bands for Long Distance Communications

Critical Frequencies for International & Nationwide Communications

While VHF/UHF frequencies assist localized coordination, HF bands offer resilient long-distance interaction when transmitting information regionally and nationally, proving important amid failures cascading beyond county boundaries. This section lists the primary high-frequency channels and dial settings optimal for interstate emergency communication up to 300 miles away without infrastructure dependent on ionospheric propagation.

Priority Frequencies:

1) 3.880 MHz - Established a calling channel for linking emergency networks and coordinating a wider regional response

2) 7.230 MHz - Nationally approved frequency for interstate radio relay discussion-related incident conditions

3) 14.060 MHz - Critical channel for communicating damage assessments and resource requests between Officials and Emergency Coordination Centers addressing state-level demands

4) 18.160 MHz - Key bandwidth for public relief groups verifying disaster information and supporting deployment decisions of reaction teams

5) 28.400 MHz Technician Class band ideal for general notice traffic over 100-mile distances

Emergency Activations:

Groups like ARES initiate coordinated emergency nets on frequencies like 3.880 MHz mobilizing formal relay links amongst operators circumventing broken infrastructure. Traffic embraces needs assessments, resource requests, and personnel dispatch. Nets manage contact ensuring opportunities.

Ionospheric Conditions, Propagation, and Band Plans

To optimally leverage High Frequency (HF) radio contact spanning hundreds of miles and infrastructure, preppers must grasp the atmospheric dynamics influencing signal propagation at different frequencies along with band planning conventions determining optimal channel selection by season and sunspot activity.

Ionospheric Propagation Primer

- Earth's ionosphere layer, commencing 50 miles up, provides skywave refraction allowing messages to reach worldwide locations in optimal conditions by 'bouncing' between Earth ground waves and ionization layers. Varying sunspot activity directly affects propagation consistency through disruption.

- Low bands between 1.8mhz to 10mhz nocturnal signals refract best for regional communications under 500 km. 40 to 10-meter wavelengths sustain dependable groundwave for < 50 miles sans ionospheric bounce.

- Current solar activity charts define the correct frequency, and D layer density allowing wavelength penetration to be crucial. High polarity periods favor lower frequencies. Low prefers higher.

Band Planning Dynamics

- Channels plan around fluctuating properties - 40m targets night/Evening transmissions; 30m daytime/afternoon; 20m daytime/critical when propagation is low globally; 10m open sporadically amid severe solar activity only.

- Digital technologies increasingly relay traffic when voice circumstances are inadequate across bands

- Alternating higher and lower attempt contact as situations morph until signals stabilize the channel

Simply, as sunspot cycles move, ionosphere density fluxes, determining which HF band best permits skip for distances sought daily. Adapting the channel accordingly permits decent communication amid atmospheric variability.

Message Handling, Logging, and Net Protocols

Formal procedures governing triggering emergency HF nets transfer crucial communications between several relay operators increasing collective support and transmitting vital intelligence assisting reaction. This closing section discusses mechanisms assuring reliable information transmission.

Calling Activation: Initiated by designated net control station broadcasting on specified net frequency for participants to acknowledge as check-ins describing functional capabilities and location to compile support resources in the emergency bank.

Message Handling: Standardized formatting accepted - Precise facts on Origin,

Destination/Requester, Priority Designation (Immediate, Urgent, etc), Related Incident Identifier (if applicable), Message Text or Content coupled with Handling Instructions like requested response timescale. This facilitates quick comprehension.

Logging: Managed chronologically by allocated sequence numbers tied to messages linking entries like an indexed emergency inbox for reference along with date-time received, station originating, and any special notes like unreceived responses.

Discipline Guidelines: Participants are granted transmit authority by net control based on traffic priority levels. Stations not transmitting assume monitoring duty copying in real-time. Overall etiquette remains calm and professional. Best operators display military-style discipline maximizing group effectiveness.

In summary, while chaotic disasters ignite around stations locally, well-managed HF emergency nets marshaling region-wide support embrace orderly procedures ensuring accurate interstate info flows are minimally distorted by unreliability or misunderstanding given infrastructure uncertainties during a crisis. The protocol indicates first-line defense guarding safety.

Chapter Seven

Portable and Mobile Setup Options

Field Radios, Go Kits, Deploying Temporary Repeaters & Antenna Solutions

Emergencies typically develop away from home stations wanting portable solutions created onsite issue locales transmitting traffic and coordinating response providers Exactly where crises emerge. This section describes ideal portable packages, temporary antenna field deployment combined with repeater insertion or interlinking bridging operability gaps when fixed infrastructure fails.

Rapidly Deployed Field Radios: Rugged, small transceivers prove critical for forward emergency communicators assessing circumstances, damage, and urgent demands directly. Waterproof handhelds like Yaesu FT-60R support 5-watt VHF/UHF power handling locally. QRP CW/Data capable radios featuring waterfall SDRs scan larger bandwidths. Handheld satellite rigs also bridge regional traffic. Go Box Solutions organizes several rigs.

Failsafe Power Sources Portable:
Goal Zero Yeti series lithium power systems now enable indefinite, continuous 50-100w solar recharge capacity allowing most 100w HF radios to run 24/7. Bioenno, Powertraveller packs offer comparable capabilities and charge handhelds indefinitely. Solar remains scalable under sustained loads.

Tactical Antennas In Field Without Structures:
NVIS wire emergency antennas stretch between trees/poles utilizing insulators providing HF regional comms where towers/buildings are

unavailable. Roll-up 2m/440mhz stainless pole antennas deploy locally using lineman straps. Dual-band tape measure Yagis boost signal gain hitting distant towers/repeaters.

Repeater Interlinking For Extended Range: Temporary crossband repeaters carried to incident frontlines using external duplexers and amplified antennas interlink into wider reaching VHF/UHF infrastructure spanning operational range for miles where only HD radios prevail otherwise limiting local comms across rolling terrain with groundwave only.

Mobile Installation in Vehicles, Emergency Power Options

Beyond individual grab-and-go field radios, preppers install strong equipment suites permanently placed in rescue/support vehicles preserving communications indefinitely when despatched to wrecked zones post-disaster to convey crucial damage assessments assisting in the deployment of material aid.

Optimized Mobile Radio Sets: Dual configuration becomes important - separating standalone VHF/UHF rig for local ground crews from dedicated HF all-mode transceiver sustaining regional backhaul links. Yaesu FTM-400XDDR runs on 45 watts hitting local repeaters reliably. Kenwood TS-480SAT delivers 100w HF and 6m reach critical when infrastructure problems separate towns.

Rugged Vehicle Antenna Solutions: All antennas must resist highway speeds and severe weather. Multi-band NMO mounts with spring bases like Comet SBB-5NMO HF+ 6M/2M/440mhz sustain wind/ice stresses using bonded nylon glass body mounts. Simple NMO stud adapters offer optimized antenna selection. RG-8X coax handles 100w.

Sustaining Reliable Mobile Power: An Auxiliary dual battery setup isolates operator gear from vehicle starting functions using 100ah Amp Hour AGM deep cycle battery reserve capacity powering communications loads exclusively via 1500w inverter connected equipment plugging into household style outlets and USB ports built-in. Dual 15w solar panels top out over hours without the need for shore power.

Tactical Call Signs, Operation Protocols for Displaced Hams

Adopting suitable call sign naming conventions & transmission protocols alongside common MARS/EMCOM terminology allows licensed operators constructing temporary stations in devastated areas to be identified by role & integrate communications supporting authorities safeguarding the region.

Tactical Call Sign Assignment: Stations specify served agency before suffix:
- ARC for NGOs like the Red Cross
- ARES for Amateur Radio relay volunteers
- DHS for DHS tactical teams(broadcasts encrypted normally)
- LOCAL NAME-COM identifying the incident area

Traffic Priority Identification: First section of call before message suggesting urgency level so all listening focus resources react accordingly:
- MEDICAL EMERGENCY
- SAFETY ALERT - RESOURCE REQUEST etc

Disciplined Channel Sharing: All operators respect 1.5-second intervals before reacting to prevent doubled transmissions. Nonessential stations stay in receive-only monitoring positions unless traffic justifies, cooperating with Net Control assignments.

Use of Standard Terminology: Common MARS, EMT, and Fire Agency terminology defines situations, procedures, and locations universally across response teams promoting inter-agency interoperability.

Adherence to protocols orders shared channels maximizing collaborative crisis response and boosting public wellbeing when afflicted regions sustain lengthy infrastructure outages. Now let's study sustaining mission-critical equipment operation long-term off-grid...

Chapter Eight

Emergency Power Systems & Managing Battery Banks

Solar, Generators, Charge Controllers - Oh My!

With grid infrastructure disabled indefinitely after catastrophic catastrophes, amateur radio networks must retain failsafe electricity for vital communications technology using renewable backup sources like solar arrays, generators, and battery banks giving continuous, reliable 12v DC current.

Effectively leveraging these alternatives eliminates preventable equipment failures needlessly endangering public safety when disasters severely destroy regional power supplies. This section outlines appropriate emergency power combinations ensuring resilience.

Solar Advantages: - Direct-to-DC charging requires just controllers safeguarding battery lifespans
- Panels rated 100 to 400 watts successfully run most HF/VHF/UHF base stations
- Peak sun hours determine real production levels - 4-8 hours for 100w solar
- MPPT vs PWM controllers balance performance, cost for loads

Quality Generators:
- Dual fuel propane/gasoline tri-fuel Honda EU7000is shows the gold standard 0 - 7-10kW ranges handle multiple radio setups
- Key for lengthy cloudy periods limiting solar recharge

Integration - Stacked Capabilities: - Chain components increasing production: 400w panels > 200ah batteries > 2kw inverter > 100w transmitter - Retain baseline generator backup given solar reliability uncertainties

Wiring Solutions for Off-Grid Shack Power

Safely channeling electricity from outside generation interior stations entails correctly gauged wires and connections eliminating fire dangers or equipment damage when big loads are operated constantly.

Critical Path Components:

1. Marine-rated battery boxes fastening components externally

2. Oversized 00 gauge power wire conducting high-watt loads

3. Large DC load disconnects before entering buildings

4. Copper bus bars transporting high amps within stations

5. Multiple filtered power pole receptacles station gear plugs into

6. Mid-line fuses protecting over current surges

7. Lightning/EMP-proof Faraday shielding and grounding

Wire Gauge Capacity Reference:

00 gauge - 300amps

0 gauge - 200amps

4 gauge - 100amps

10 gauge - 50amps

With capacities recognized, designing home run wiring from battery banks via control rooms eases power provision pressures avoiding overloaded conditions causing fires given long-term sustained emergency radio ops.

Maintaining, Storing, and Charging Deep Cycle Station Batteries

Reliable off-grid power begins with sturdy deep cycle batteries safely holding electricity generated from solar arrays and generators, available for immediate distribution powering transceivers as crises unfold. Careful selection, charging, and maintenance assure longevity.

Optimal Battery Selection:

- 12v Deep Cycle Lead Acid: Traditional and inexpensive alternatives still dependably provide DC stations for years despite lithium improvements when properly maintained. Common sizes - 35ah / 75ah / 125ah

- 12v Lithium Iron: Lightweight, environmentally-sealed LiFePO4 packs survive 3000+ cycles. Operate in any orientation. 100ah packets suitable for grab-and-go kits. But prevents freezing, unlike lead acid.

Battery Charging:

- Never pull below 50% charge avoiding rapid capacity deterioration over time through deep discharge damage. Solar frequently brings to 100%.
- Equalization modes on sophisticated charge controllers bolster cells' voltage balancing
- Timer settings avoid overcharging if unattended for lengthy durations

Battery Care:

- Use a hydrometer to routinely monitor acid-based voltage levels finding deteriorating cells

- Clean rusted terminal connections impeding performance
- Store indoors away from elements in clean, dry locations
- Label Each battery cabling by ID for convenient load testing

Carefully selecting long-life batteries for the frequent deep cycling communication loads demand combined with proper sizing, charging and testing of designed-to-fail cells offers the best probability stations stay continuously powered facilitating community support through extended disruptions as infrastructure rebuild lags.

Chapter Nine

Antennas and Propagation for Emergency Communications

NVIS, Dipoles, Verticals, Beams: Picking Optimal Antennas

An amateur radio station is only as effective as its antenna design permitting broadcast signals to reach intended recipients across specified distances during infrastructure outages. This important component transforms stable power sources and robust transceivers into life-saving capabilities when disasters occur. By matching propagation needs to purpose-built antenna designs correctly, even lower wattage radios reliably reach 50+ mile radii in hard terrain securing the continuity of communications and protecting communities when catastrophic catastrophes isolate regions.

NVIS Antennas: Near Vertical Incidence Skywave wire antennas excel joining local stations up to 400 miles apart without repeaters at heights allowing skywave signals refracting off the ionosphere to come back down near operators enhancing groundwave overlap. Ideal for rural inland county-level coverage. Dipoles set below 10 mhz perform best taking advantage of nocturnal D layer refraction.

Beam Antennas: Directional Yagi and loop beams direct signal strength exactly where needed most pointing at distant repeaters or base installations improving both send and receive district-level range capabilities when vast area emergency coordination is vital. High-gain driven elements inside metal boom housing assist VHF/UHF signals crossing gaps.

Omni-Directional Antennas: When wide area 360-degree transmission and reception are sought for general area simplex operations, then omnidirectional dipole, disc one, and ground plane verticals prove best with no manual pointing or adjustment required. VHF/UHF coverage up to 20 miles is feasible.

The finest emergency communicators possess antenna variety allowing adaptive setup for the environment and bands active as disasters develop geographically. Let's analyze installation procedures and best placement amplifying settings.

Installing Your Antenna and Using Analyzers

With ideal antenna types now matched with frequency bands and specific propagation modes specified, we pay attention to physically elevating equipment for establishing both fixed and temporary solutions enabling emergency networks as regional infrastructure fails.

Base Station Installation: Fixed external tower structures:
- Roof tripod mounts fasten to chimneys and peaks via adjustable collars without penetrating shingles using ratchet booms ranging up to 40 feet safely.

- Freestanding crank-up towers permanently positioned safely on ground planes with solid concrete-based counterweight designs stretch telescoping triangle masts to 60-foot heights capable of supporting massive multi-band directional beam arrays by rotor motor controls.

Analyzer Usage Vital Post-Install: Following safe installations, antenna analyzers then extensively evaluate designs throughout planned frequency ranges guaranteeing good SWR impedance matches free of defects degrading propagation. Analyzers discover dips and spikes informing corrections assuring efficacy in hitting goal distances during disasters.

Common analyzer reference models: - Comet CAA-500 thoroughly checks antennas 10hz to 525 mhz finding flaws
- nanoVNA V2 contacts 137mhz scanning wide spectrum performance in sharp detail

Consistency in obvious propagation patterns actually demonstrates antenna assemblies are refined for mission-important emergency usage when crises unfold regionally.

Managing Antennas Off-Grid with Rotators & Tuners

In chaotic catastrophic occurrences when infrastructure damage fluctuates daily, emergency networks must stay adaptable diverting antennas when VHF/UHF repeaters/services shift or HF NVIS nocturnal ideal band circumstances vary. This capability depends on redundant rotator placement controls with onboard adjustable components manually matching bands supported when stations activate serving devastated districts.

Rotator Controllers: Compact embedded and indoor display controllers rotate outdoor antenna arrays remotely via Molex wiring and alignment motors calibrated shifting 360 degrees targeting key repeaters, net control stations, or skywave headings by operator adjusting pitch and azimuth settings gradually to optimize signal clarity relaying emergency traffic as needed. Quality controllers draw minimal 12v current suited for continuous off-grid usage.

Tuner Units In Shack: Manually adjustable impedance tuners placed inline between transmitters and antenna feedlines adapt high frequency configurations over complex loads above 500 khz increasing power transfer efficiency and SW ratios allowing field operators continuity contacting regionally dispersed aid forces and officials daily across emergency networks as propagation conditions morph in disaster wake. Tuners are especially critical where NVIS nightly HF skywave transmissions key coordinate relief.

Careful orientation adjustments bridging damaged areas off-grid in combination with adaptive tuning backup stranded stations remain information lifelines for communities during outages. We shall now explore tying all this capability into complete readiness strategies.

Chapter Ten

Finalizing Your Capabilities & Next Level Preparedness

Failover, Redundant Gear, and Rapid Deployments

With primary comms hardware selected for bases and field stations, preppers build resilience via failover redundancy across every component protecting collective emergency coordination when catastrophic crises occur and limiting certain frequencies, gear, or power infrastructure for periods of time. Backups multiply flexibility keeping communities connected.

Transceivers when Lifelines: Install clearly labeled grab-and-go containers stocked with handheld

5-watt transceivers like Yaesu FT70DRs covering important amateur bands vulnerable to destruction or crowding when large disasters unfold. Having gear disseminated to rally locations allows continuity when shacks overrun. Waterproof cases secure electronics for extended periods fitted go-packs connect.

Power Stripped Bare? Rotate Renewables:
With battery banks depleted by days of 100% radio duty cycles, immediately reposition portable solar panels southern tilting for max sun exposure rates to resume trickle charging 12v lifepo4 recovering exhausted deep cycles to 50% usefulness revolving, permitting transceiver power continuity. Panels as tiny as 10 watts sustain minimum comms working.

Antennas Destroyed? Improvise Field Expedient Replacements: Where fiberglass/metal permanently installed antennas become damaged by violent weather, quickly construct temporary wire dipoles hoisting inverted V configurations between structures which allow makeshift connectivity restored despite compromised permanent antennas remaining inoperable for long periods post-disaster until extensive repairs are possible.

Communications Backup Plans: GMRS, FRS, MURS, CB Radio, SSD

When ham bands congest or fail altogether under huge usage loads with local repeaters down during catastrophic disasters, always stay capable of reaching crucial friends via multiple consumer frequency alternatives as contingencies mentioned below.

GMRS/FRS Radios - Portable user-friendly band radiating 5-8 kilometers without infrastructure suitable for neighborhood watches and local reaction teams handling events like fallen lines, gas spills, and light SAR. Wouxun KG805G models include GMRS transmit with FRS receive monitoring, ideal for groups.

MURS Band - Widely neglected VHF shared public frequency pool about 151-154 MHz used for short-range peer connections up to 2 miles around incidents complementing phones locally with adequate terrain like RV and boaters use more recreationally. Useful backup frequency source as HAM options narrow.

CB Band Channel 9 - Monitored by professional drivers and ECCs voluntarily, Channel 9 emergency frequency pools trucks for mass transit coordination when roadways are obstructed plus disseminates roadway hazard information to vehicles despite regional disruptions.

SSD Texting - Satellite texting devices like inReach and Spot utilize orbital networks sharing 160-character messages transmitting crucial intelligence like geo coordinates, urgent situations and operational conditions from remote blocked places back to infrastructure on outside of disasters. Useful for telemetry and confirmations when other backups are insufficient.

Staying Active in the Hobby and Community

Beyond simply accumulating functional capabilities via amateur radio assets actively preparing for infrastructure instability, genuine resilience derives from relationships, knowledge, and realistic skill rehearsal only possible by staying dynamically involved with the nationwide community of peers organizing around emergency response volunteer locally.

Attending Area Club Meetings - Consistently connecting with local club leaders and members in monthly sessions creates links tightening common preparedness to mobilize when regional catastrophes strike close home. Evolving best practice ideas penetrate groups during this face time and fellowship.

Assisting Regional Nets - Participating in weekly training nets hosted by volunteers familiarizes processes and numerous EmComm disciplines such as traffic handling, logging, and tactical communications. Instilling readiness, comfort managing genuine situations converting information into action requires rapid thinking as scenarios shape-shift.

Supporting License Exams - As local radio clubs host exam sessions converting newcomers into licensed operators, serving as proctors and technical advisors aid beginners, expanding ranks of available emergency communicators with legal frequencies afforded transmit privileges during incidents when additional relay volunteers essential facilitating ER responses.

Fabricating Gear/3D Printing - Pursuing technical self-sufficiency skills like assembling antennas from scratch using copper tubing and wire alongside CAD design and 3D printing of radio accessories like signaling lanterns or mounting hardware tackles customization barriers when commercial options are lagging.

In Summary, Well-rounded community engagement on numerous fronts beyond merely purchasing ruggedized equipment assures hams aiding regionally enduring crises know how to deploy their capabilities to defend civilians most effectively till stability is restored from turmoil.

Exclusive Bonus

20 Non-Ham Communication Alternatives for Preppers

While ham radio reigns supreme in emergency communication circles, a prepper's toolbox shouldn't be confined to just one gear. Here, we investigate 20 alternate communication techniques for when the grid goes down and the silence screams:

1. Walkie-Talkies

Walkie-talkies allow short-range handheld communication without having a large infrastructure. High-quality models have a range of several kilometers for coordinating local activities.

2. CB Radio

CB radio offers communication up to 10-15 kilometers of line-of-sight. It continues working during power outages and crises where cell towers are down. Channels 9 and 19 are emergency frequencies monitored by many users.

3. FRS/GMRS Radios

Family Radio Service (FRS) and General Mobile Radio Service (GMRS) radios permit short-range communication up to several kilometers without needing a license. Some versions communicate directly, while others require repeater towers.

4. Satellite Phones

Satellite phones connect directly to satellites rather than ground-based cell towers, allowing communication globally during calamities. Some handheld devices like inReach additionally provide GPS positions and SOS signaling.

5. Emergency Beacons

Personal locator beacons (PLBs) and EPIRB boat beacons broadcast GPS location coordinates to satellites, alerting emergency rescuers during wilderness rescues or maritime events when other communications fail.

6. High-Frequency (HF) Single Side Band Radios

HF SSB radios carry voice and data over great distances by reflecting signals off the ionosphere up to 1,500 miles away without repeaters. Frequencies between 1.6-30 MHz are frequent.

7. Airband Scanners

Programmable airband aviation radios pick up Emergency Locator Transmitter distress signals and ease listening to National Weather Service broadcasts for advanced disaster warnings when available.

8. Text Messaging Apps

Peer-to-peer messaging apps like Firechat, Bridgefy, and Briar link directly to adjacent phones without using cell towers, facilitating communications during mass disaster events near users.

9. Wireless Emergency Alerts (WEA)

Cell carriers broadcast location-based WEA text messages to WEA-capable phones during emergencies, even when towers fail. Useful for mass evacuation and vital directives.

10. Infrared Beam Communication

Infrared modules that broadcast modulated beams vast distances provide basic low-bandwidth communications impervious to radio interference allowing reasonably clandestine usage between operators.

11. Remote Radio Link Extenders

Strategically positioned portable repeater stations located in high locations can transmit signals between topographical obstacles like mountains and valleys in vulnerable areas that lack a direct line of sight.

12. Emergency Notification Sirens/Speakers

Battery or hand-cranked powered public address speaker devices installed across neighborhoods restore localized transmission of emergency messages to inhabitants using mic input or pre-recorded voice memos.

13. Fiber Optic Mesh Networks

Resilient mesh-like high-speed fiber optic cable network infrastructure connects decentralized regional communication centers providing administration facilities with stable connectivity. Expensive dense installation but unbreakable virtually by disasters once completed.

14. Delivery or Courier Services

When electronic communications are unavailable, pre-established contingency courier services safely send foot, bike, and car messengers transporting written documents and packages along routes connecting network members to reestablish crucial channels for coordinating.

15. Modified Vehicle Horn Messaging

Certain patterns of car/truck horn sounds could conceivably send simple pre-planned messages to local community members within earshot like all-clear alerts or pleas for medical aid when hazard limits foot delivery.

16. Improvised Signal Mirrors

Crude but effective, DIY handheld signal mirrors created from polished metal or foil when tilted properly towards sunlight reflect dazzling beam flashes visible for miles line of sight - a durable direct visual communications conduit resistant to device failure or interference.

17. Marker Smoke Signals

Various hues of smoke generated from exact component combinations burned for visibility (avoiding toxicity) send distinct messages observable regionally when populations lack electronics but yet require coordination across distances greater than line-of-sight optics range alone.

18. Signal Flags Like maritime applications but ashore, dividing solid color flags between partner locations combined visually in distinct orientation patterns by participants relay coded messages rapidly over a number of miles conveying key updates like resource requests allowing quicker large-scale coordination.

19. Pen Flares and Signal Rockets

Military-type pyrotechnic signaling devices launch phosphorus-bright projectiles hundreds of feet emitting distinctive patterns sending crucial messages for 5+ miles over barriers - robust last-resort solution entirely analog and interference-proof.

20. Radio/Television Broadcasts

Restoring even primitive AM/FM/TV/Shortwave broadcasts via adapted transmission sites with studios in secure places gives crucial one-way emergency communication and morale-lifting public information channels to isolated communities when two-way choices all collapse post-disaster.

www.ingramcontent.com/pod-product-compliance
Lightning Source LLC
Chambersburg PA
CBHW071052290526
45795CB00004B/1450